Charles Darwin

Anita Croy

CRABTREE

PUBLISHING COMPANY

WWW.CRABTREEBOOKS.COM

CRABTREE
PUBLISHING COMPANY
WWW.CRABTREEBOOKS.COM

Author: Anita Croy
Editors: Sarah Eason, Melissa Boyd, Ellen Rodger
Proofreader and indexer: Jennifer Sanderson
Proofreader: Wendy Scavuzzo
Editorial director: Kathy Middleton
Design: Paul Myerscough and Lynne Lennon
Photo research: Rachel Blount
Print coordinator: Katherine Berti

Written, developed, and produced by Calcium

Photo Credits:
t=Top, c=Center, b=Bottom, l= Left, r=Right

Inside: Shutterstock: Ase: p. 52; Beboy: p. 12; Samuel Borges Photography: p. 30; Volodymyr Burdiak: p. 49; Henryk Ditze: p. 26; Dreamansions: p. 42; Everett Historical: p. 48; Fotos593: p. 29; GregGrabowski: p. 59; Rene Holtslag: p. 19; Jiri Hrebicek: p. 31t; ImageFlow: p. 58; Kjersti Joergensen: p. 36; Jopelka: p. 39t; Jorisvo: p. 7r; Juefraphoto: p. 45; Olga Korneeva: pp. 10t, 10c, 11tr, 11br; Eileen Kumpf: p. 15; Lynea: pp. 34tr, 34tr, 35t, 35c; Discover Marco: p. 28; Wilfred Marissen: p. 37cl; Morphart Creation: p. 41b; Lucas Correa Pacheco: p. 25c; Paleontologist Natural: p. 27; Ksenia Ragozina: p. 51; Rudenkois: p. 16; Beth Ruggiero-York: p. 44; Aleksey Stemmer: p. 61; Stubblefield Photography: p. 37t; KASIRA SUDA: p. 9b; Emre Terim: p. 54; Travellight: p. 9t; Sergey Uryadnikov: p. 32; Bjoern Wylezich: p. 11cr; YaBarsArt: p. 26l; Wikimedia Commons: pp. 35r, 55; Julia Margaret Cameron/Wellcome Images: p. 47; Jens L. Franzen, Philip D. Gingerich, Jörg Habersetzer1, Jørn H. Hurum, Wighart von Koenigswald, B. Holly Smith: p. 50; John Gould: p. 40; Ernst Heinrich Haeckel: p. 60; Lewis Wickes Hine: p. 57; Lock & Whitfield/Wellcome Images: p. 46; London Stereoscopic & Photographic Company: p. 39c; Charles Lyell: p. 21; T. H. Maguire/Wellcome Images: p. 10b; Henry Maull & John Fox: p. 4; R. T. Pritchett: p. 22; George Richmond: p. 34l; Ellen Sharples: p. 6; J. R. Smith/Wellcome Images: p. 20; The zoology of the voyage of H.M.S. Beagle: p. 38; Walker & Cockerell: p. 37cr; Dr. Wallich, Oscar Gustave Rejlander and Guillaume Duchenne: p. 41r; Friedrich Georg Weitsch: p. 17; Wellcome Images: pp. 24, 56; Joseph Wright of Derby: p. 14.

Library and Archives Canada Cataloguing in Publication

Title: Charles Darwin / Anita Croy.
Names: Croy, Anita, author.
Description: Series statement: Scientists who changed the world |
  Includes index.
Identifiers: Canadiana (print) 20200225820 |
  Canadiana (ebook) 20200225901 |
  ISBN 9780778782186 (hardcover) |
  ISBN 9780778782247 (softcover) |
  ISBN 9781427126108 (HTML)
Subjects: LCSH: Darwin, Charles, 1809-1882—Juvenile literature. |
  LCSH: Biologists—Biography—Juvenile literature. | LCSH:
  Naturalists—Biography—Juvenile literature. | LCSH: Evolution
  (Biology)—History—Juvenile literature. | LCGFT: Biographies.
Classification: LCC QH31.D2 C76 2021 | DDC j576.8/2092—dc23

Library of Congress Cataloging-in-Publication Data

Names: Croy, Anita, author.
Title: Charles Darwin / Anita Croy.
Description: New York : Crabtree Publishing Company, [2021] |
  Series: Scientists who changed the world | Includes index.
Identifiers: LCCN 2020017177 (print) |
  LCCN 2020017178 (ebook) |
  ISBN 9780778782186 (hardcover) |
  ISBN 9780778782247 (paperback) |
  ISBN 9781427126108 (ebook)
Subjects: LCSH: Darwin, Charles, 1809-1882--Juvenile literature. |
  Naturalists--England--Biography--Juvenile literature.
Classification: LCC QH31.D2 C67 2021 (print) |
  LCC QH31.D2 (ebook) | DDC 576.8/2092 [B]--dc23
LC record available at https://lccn.loc.gov/2020017177
LC ebook record available at https://lccn.loc.gov/2020017178

Crabtree Publishing Company
www.crabtreebooks.com   1-800-387-7650

Printed in the U.S.A./082020/CG20200601

Published in Canada
Crabtree Publishing
616 Welland Ave.
St. Catharines, Ontario
L2M 5V6

Published in the United States
Crabtree Publishing
347 Fifth Ave
Suite 1402-145
New York, NY 10016

Published in the United Kingdom
Crabtree Publishing
Maritime House
Basin Road North, Hove
BN41 1WR

Published in Australia
Crabtree Publishing
3 Charles Street
Coburg North
VIC, 3058

Contents

CHAPTER 1

Charles Darwin, photographed here in about 1854, challenged people's long-held beliefs about how Earth and everything in it was created.

CHARLES DARWIN
Biography

Born: February 12, 1809

Place of birth: Shrewsbury, England

Mother: Susannah Wedgwood Darwin

Father: Robert Waring Darwin

Famous for: His **theory** of **evolution**, which was based on the idea of **natural selection**. The modern theory of evolution is based upon it.

How he changed the world: Darwin's theory of evolution changed how people saw the world and their place in it. He challenged the long-held belief that God had created every living thing in its current shape. He showed how he believed that animals had evolved, or changed over time.

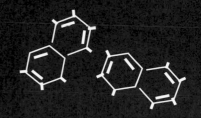

Charles Darwin described how animals that are better ADAPTED to their environment tend to SURVIVE and have more OFFSPRING. He named this idea NATURAL SELECTION.

Charles Darwin.

AN ENQUIRING MIND

Charles Darwin was born into a wealthy family in the small town of Shrewsbury, in rural England, in 1809. He was the fifth of six children. His father, Robert, had followed in his own father's footsteps and was a successful doctor. Darwin's mother, Susannah, came from the famous Wedgwood family, which owned china factories. Charles's happy childhood was cut short by his mother's death when he was just eight years old. After that, he was raised by his three older sisters.

Charles was sent away to school at age nine.

Growing up in his family's large house in Shrewsbury, Darwin loved hunting, as well as collecting all sorts of things: pebbles, plants, insects, and bird eggs. He was always in the family's garden observing nature. His father loved plants and collected rare **specimens**, which he planted on the grounds of the house. One time, his father gave Charles the job of counting all the peony flowers (384) in the garden. Charles and his older brother Erasmus, whom he worshipped, had their own chemistry laboratory in the house where they carried out experiments. Charles loved experimenting so much that his friends named him "Gas."

In this portrait showing Charles, at age seven, with his sister Catherine, he holds a plant to show his interest in nature.

Ideas that changed the world

In his book, The Descent of Man, *Darwin wrote:"I have given the evidence to the best of my ability; and we must acknowledge, as it seems to me, that man with all his noble qualities…still bears in his bodily frame the indelible stamp of his lowly **origin**."*

Exploring the ideas

In 1871, long after he had published *On the Origin of **Species*** (1859), Charles Darwin published *The Descent of Man*. The work was the result of his lifetime's study. Darwin argued that human beings were part of the animal kingdom and had evolved just like any other species. This link between humans and animals shocked many people, but it illustrated how evolution worked: over millions of years, humans had evolved into the beings we now are. Yet, when Darwin was writing, many people believed the Bible was an exact account of Earth's history. The Bible said that God had created humans and animals just as they were. Over the centuries, many thinkers had questioned that idea because they could see that the world changed over time, but since such evidence did not fit in with Bible teachings, it was usually dismissed.

The Church taught that God had created humans by making Adam and Eve.

HISTORY'S STORY

Charles grew up knowing his grandfather's ideas about evolution. His grandfather, the physician Erasmus Darwin (1731–1802), wrote the book *Zoonomia* (1794–1796), in which he suggested that species changed over time because of **sexual selection**, which is how animals choose a **mate**. However, he did not have proof to back his theories.

A POOR STUDENT

When he was just 16, Charles accompanied his older brother Erasmus to the leading medical school in Great Britain, the University of Edinburgh. Their father decided that both boys would follow in his and his father's footsteps. The only problem with his plan was that Charles hated medicine—especially surgery, which was then done without effective **anesthetic**.

In the early 1800s, medicine could be a **barbaric** profession. Surgeons operated in filthy robes because no one had yet made the connection between disease and hygiene, or cleanliness, and surgical instruments were basic. After watching an operation on a child, Darwin vowed he would never go into an operating room again. He knew that medicine was not for him, but it was two years before he got the courage to tell his father that he wanted to stop studying.

Charles hated medicine—especially surgery.

Radical ideas

The University of Edinburgh had a reputation for being home to many students with radical, or extreme, views. Unlike the universities of Oxford and Cambridge, Edinburgh admitted "dissenters," or people who did not agree with accepted views of Christianity. In Edinburgh, Darwin joined the Plinian Society, named for the ancient Roman **natural historian** Pliny the Elder. Members of the society were men who were interested in natural history, which covers the nature of all living things. They met every week to discuss research papers on different topics. Many of these papers openly challenged the religious views of the day. Although Darwin failed to become a physician, his two years in Edinburgh were not wasted: he studied **natural philosophy**, went to talks, and collected specimens.

His true passion

Darwin's father decided Charles needed a career, so he sent him to study divinity, or religion, at the University of Cambridge. He wanted his son to become a priest, which was considered a very good profession. In 1828, Darwin arrived at Christ's College in Cambridge. The move transformed his life. While he studied hard enough to graduate in 1831, he spent as much time as he could with his friends, including the **geologist** Adam Sedgwick (1785–1873). Surrounded by likeminded people, Darwin's interest in animals, plants, and rocks continued to grow.

The beetle craze

During the 1820s, a craze for collecting beetles swept through England. Darwin soon joined in. With his cousin, William Darwin Fox, he spent a lot of his time "beetling." Collectors paid well for rare species. One day, Darwin discovered two rare beetles and grabbed one in each hand. Before he could put them in his collecting box, he spotted an even rarer third species. In his excitement to claim all three, he popped one of the beetles into his mouth to free up a hand. This was a big mistake. The beetle was the dangerous bombardier species. It sprayed his mouth with burning fluid!

Christ's College was founded in Cambridge in 1437.

When they are attacked, bombardier beetles spray chemicals with a loud pop.

STUDYING THE WORLD

Darwin spent his time in Cambridge studying—but not studying divinity as his father had intended. Surrounded by fascinating people, Darwin explored many new subjects. As well as collecting beetles, he studied a group of animals called **invertebrates** and the scientific **classification** of plants. He also made friendships that would dramatically alter the course of his life.

At Cambridge, Darwin particularly enjoyed the lectures of the **pioneering** professor of **botany** John Henslow (1796–1861). Henslow's lectures were among the most popular classes at Cambridge. Forgetting his divinity lectures, Charles sat in on many of the professor's classes and even joined his field trips. Henslow did not believe in spoon-feeding his students. Instead, he taught them to go out and observe nature for themselves. Henslow taught Darwin how to look at nature with a scientist's eye, carefully noting detail. Darwin was beginning to realize that science was not to be found in dusty classrooms, but instead in the world around him. Later in his life, Darwin described his friendship with Henslow as the most important in his life.

John Henslow spent much of his time collecting and classifying plants.

A fashion for classification

Darwin's studies at Cambridge reflected a trend for what in the early 1800s was called natural philosophy, but is today known as natural history. The trend was based partly on the importance that people of the time put on taxonomy, or classifying plants and animals in a way that showed the relationship between them. Most classification was based on a system of species and genus created by the Swedish botanist Carl Linnaeus (1707–1788).

Learning to love geology

In Edinburgh, Darwin had been bored by **geology**. However, once he met Adam Sedgwick at Cambridge, he realized how fascinating rocks could be and how much information they held. Not only did rocks come in all kinds of different colors. They also contained **fossils**, which geologists were beginning to realize held many secrets about ancient animals and plants.

Fossils are the preserved remains of ancient life forms, such as fish, that no longer exist.

HISTORY'S STORY

The botanist Carl Linnaeus gave every species two Latin names: a genus name and a species name. A species, such as lions, is a group of living things that look and behave similarly. A genus contains species that are closely related to each other, such as all the big cats that can roar. Linnaeus's system is still used today.

11

CHAPTER 2

Geologists study volcanoes to see how Earth's rocks are formed.

SCIENCE AND RELIGION
Background

Questions for 19th-century natural philosophers included:

What relationship did fossils have to modern plants and animals?

What was the connection, if any, between animals and humans?

Why do two children in the same family look different from their parents and each other?

How old is Earth?

Has Earth changed since the day it was created?

Were all the living things on Earth the same as the day God made them?

How do living things adapt to changing environments, such as different weather patterns or habitats?

Darwin wrote that it was ARROGANT for humans to think they were so special they could only have been created by a GOD. He said it was more HUMBLE and more ACCURATE to believe humans were created from animals.

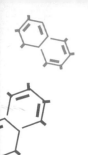

GENTLEMEN AND SCHOLARS

Charles Darwin was born at a time when most people still believed the world was the work of a **divine** being, God, who had created it just as they saw it around them. This theory fit with the Bible's Book of Genesis, which told the story of creation. However, there was a growing body of people, called natural philosophers, who were beginning to challenge this idea.

In this 1768 painting by Joseph Wright, a natural philosopher performs an experiment in which a bird dies from lack of oxygen.

The natural philosophers were mainly **gentlemen** scientists who came from wealthy families. They had the time and money to carry out investigations and experiments into areas of natural science that interested them. These men lived at an interesting moment in history. Modern science, based on careful experiments and observations, had started to take shape in the 1600s. In Darwin's day, however, science was still not taken seriously—it was not even taught as a subject in school. However, things were about to change.

In the 1800s, the **Industrial Revolution** was changing the world with new technology, while gentlemen scientists were starting to make important observations and breakthroughs. Some of their ideas were challenging the accepted view of God's role in the creation of the world. Most people were not ready for such radical ideas.

Ideas that changed the world

Darwin wrote: "But I own that I cannot see as plainly as others do, and as I should wish to do, evidence of design and beneficence [doing good] on all sides of us…I cannot persuade myself that a beneficent and omnipotent [able to do anything] God would have designedly created the Ichneumonidae with the express intention [only purpose] of their feeding within the living bodies of caterpillars, or that a cat should play with mice…"

Exploring the ideas

Darwin discussed the tiny wasps in the Ichneumonidae family, which are parasites. They live inside or on other insects, slowly destroying them. He argued that he could not see how any God would deliberately create such a cruel creature. However, he did later write that there may have been "designed laws" that set in motion the process of the world taking its current form. However, the randomness of chance, creating one result but not another, bothered him because it suggested a deep uncertainty.

Darwin wondered why a kindly God would create Ichneumonidae wasps.

HISTORY'S STORY

The biblical story of the creation of the world, told in the Book of Genesis, recounts how God created the world in six days and rested on the seventh day. In the early 1800s, most people—including Darwin—accepted this account as literally true. As Darwin studied his observations, he became less certain.

Paley's argument remained influential for the first four decades of the 1800s. He argued that God must exist because only a divine being could have created a world that was so complex. Paley argued that nature fits together perfectly, like the workings of a watch. He felt that all the animals in the world also fit together in a system far more complicated than a watch. This had to be the work of a designer. Paley explained, "That designer must have been a person. That person is God."

Paley believed a design, like a watch, must have a designer.

While he was a student at Cambridge, Darwin read Paley's *Natural Theology or Evidences of the Existence and Attributes of the Deity* (1802). In this book, Paley put forward a further argument that God had created everything: every animal was designed to fit into its environment. Darwin could see this was true, but he came to explain it in a different way.

Using reason

Another person who made an impression on Darwin was the famous mathematician and **astronomer** Sir John Herschel (1792–1871). Herschel argued that the most important aim of natural philosophy was to understand the laws of nature. He said this was best done through inductive reasoning, or using observations to draw conclusions. He contrasted this with deductive reasoning, where someone gathered observations to support what they already believed. Herschel admitted that inductive reasoning usually only led to probable conclusions, not definite conclusions. This degree of uncertainty suited a time when people were beginning to question biblical teachings.

A famous explorer

The **Prussian** explorer and scientist Alexander von Humboldt (1769–1859) was another influence on Darwin. Humboldt made a scientific expedition to Central and South America in the early 1800s. While traveling, he collected many specimens and made measurements of Earth's weather. When he returned to Europe, he spent the next 23 years writing up and publishing the **data** he had collected, which made him a celebrity in the scientific world. He also helped popularize science through his writings.

Alexander von Humboldt studied how plants lived in different climates.

UNDERSTANDING LIFE

Alexander von Humboldt's writings inspired Charles Darwin to follow in his footsteps. His friendships with John Henslow and Adam Sedgwick also encouraged him to pursue his passion for natural history. Darwin wanted scientific adventure and that could not be found in an English church.

After reading Humboldt's description of his travels to South America and to the Canary Islands in the Atlantic Ocean, Darwin decided that after graduation he wanted to go to the Canary Islands himself. He wanted to study the plants and animals Humboldt had described. He even started to learn Spanish to prepare for the trip, although he had no idea how he would get to the islands. While he waited, he decided to go on a field trip with Adam Sedgwick to improve his geology skills and gain a wider understanding of the natural world.

Darwin wanted scientific adventure and that could not be found in an English church.

While Darwin was in Wales with Sedgwick in 1831, John Henslow received a letter. His friend Captain Robert FitzRoy (1805–1865) had written to ask Henslow to accompany him on a two-year voyage to South America on the ship HMS *Beagle*, where he would draw maps for the Royal Navy. Unable to accompany his friend, Henslow suggested that Darwin would be the ideal companion.

Ideas that changed the world

*Describing the Galápagos Islands, in the Pacific Ocean, Darwin wrote: "By far the most remarkable feature in the natural history of this **archipelago**...is that the different islands...are inhabited by a different set of beings...I never dreamed that islands, about 50 or 60 miles apart...would have been differently tenanted."*

Exploring the ideas

The differences Darwin noticed between the species that lived on the different islands led him to conclude that animals adapted to fit their specific environment. Birds' beaks were shaped depending on the foods that were most easily available to them, for example. This gave Darwin important clues about how natural selection worked.

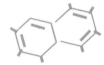

The Galápagos Islands lie roughly on the equator, off the western coast of South America.

HISTORY'S STORY

Darwin was not the only scientist who was interested in how species evolve. As early as 1801, the French **naturalist** Jean-Baptiste Lamarck (1744–1829) had developed a theory of the "Inheritance of Acquired Characteristics." He argued that, if an **organism** changed to adapt to its surroundings, these changes were passed on to its **offspring**.

LEADING NATURAL PHILOSOPHERS

Carl Linnaeus (1707–1778)

The Swede Linnaeus devised the binomial, or two-name, system of classification for living things that is still used today. He divided all life into two kingdoms—plants and animals. He then divided both into smaller and smaller categories, of which the smallest was species. Although his system is still used today, scientists now make use of seven kingdoms, including fungi and **bacteria**.

James Hutton (1726–1797)

The Scottish geologist was the first to publish the idea that the same geological processes that made Earth also continued after Earth's formation. He came to the conclusion that Earth was much older than the 6,000–10,000 years that the Bible suggested.

Erasmus Darwin (1731–1802)

Darwin's grandfather believed that life started with tiny ocean creatures that had changed over millions of years into modern life forms. He did not know how this process might have happened. He also believed man and apes were closely related. Erasmus published his ideas in poems to prevent them from being attacked as anti-Christian.

Erasmus Darwin was a physician and a leading natural philosopher.

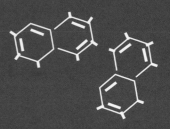

Jean-Baptiste Lamarck (1744–1829)

The Frenchman was one of the first people to propose that species evolved over a long period of time into one or more different species. Lamarck suggested that species changed as they adapted to changes in their environment.

Thomas Malthus (1766–1834)

An English **economist**, Malthus showed in his *Essay on the Principle of Population* that only the most successful of any species survived. He said that life was an ongoing struggle to live into adulthood, an idea that became known as "the survival of the fittest." The idea made an important contribution to Darwin's theory of evolution.

Charles Lyell (1797–1875)

Influenced by James Hutton, Lyell argued in *Principles of Geology* (1830–1833) that Earth was far older than anybody had thought, and that geological processes had changed it over millions of years. That suited Darwin, who needed Earth to be extremely old to fit the time scale of his theory of evolution.

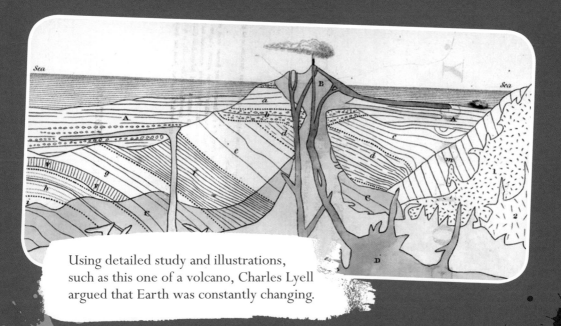

Using detailed study and illustrations, such as this one of a volcano, Charles Lyell argued that Earth was constantly changing.

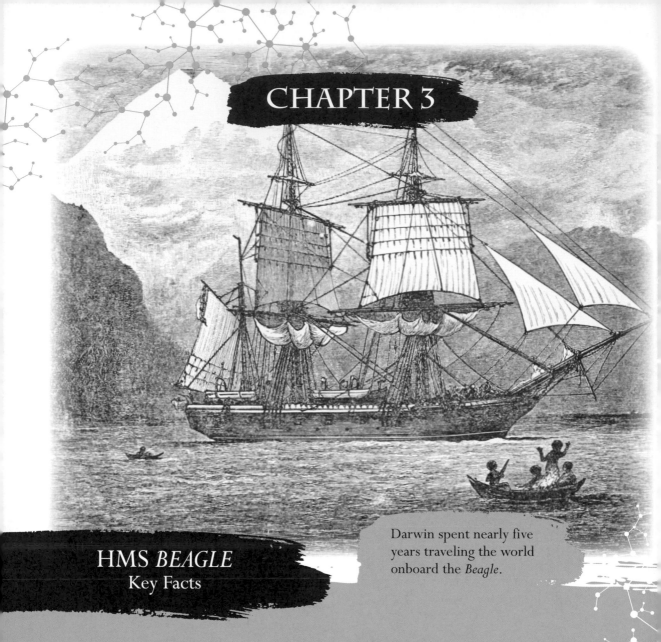

CHAPTER 3

HMS *BEAGLE*
Key Facts

Darwin spent nearly five years traveling the world onboard the *Beagle*.

HMS *Beagle*: 90 feet 4 inches (27.5 m) long, 24 feet 6 inches (7.5 m) wide

Number of crew: 74, including 10 officers and 38 seamen

Captain: Robert FitzRoy

Departed from England: December 27, 1831

Returned to England: October 2, 1836

Reason for voyage: To make maps of ocean waters around South America

Food: 6,000 cans of preserved meat

Route: Across the Atlantic Ocean via the Cape Verde Islands; around South America to the Pacific Ocean; stopping at the Galápagos Islands, Tahiti, New Zealand, Australia, Mauritius, and South Africa; return to England

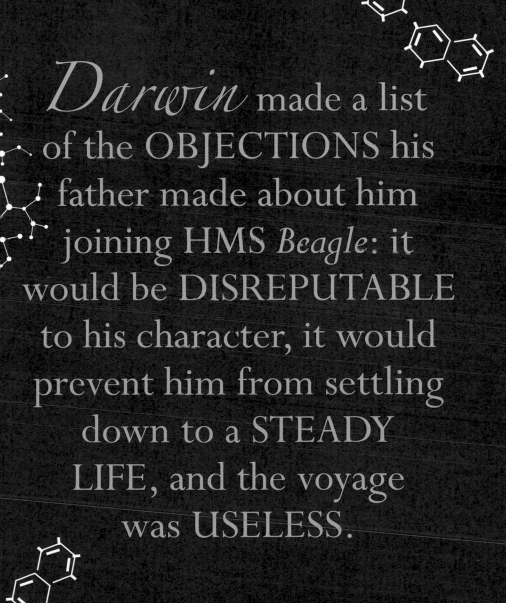

Darwin made a list of the OBJECTIONS his father made about him joining HMS *Beagle*: it would be DISREPUTABLE to his character, it would prevent him from settling down to a STEADY LIFE, and the voyage was USELESS.

AN OCEAN VOYAGE

On December 27, 1831, after several months' delay, HMS *Beagle* finally set sail from Plymouth. Onboard was Charles Darwin, who had taken all the scientific equipment he would need for a journey that was supposed to take two years. In the end, it took almost five years.

No sooner had the *Beagle* set sail, than Darwin was violently sick. He suffered from terrible seasickness for the entire voyage. More disappointment followed. The ship was not allowed to dock at the Canary Islands, off the coast of Africa, because of an outbreak of disease on the islands, so Darwin's dream of following in Humboldt's footsteps was dashed.

No sooner had the Beagle *set sail, than Darwin was violently sick.*

Darwin wrote that it was hard to believe the Fuegians were "inhabitants of the same world."

The ship arrived in Brazil on February 28, 1832. Darwin was amazed by foods and plants he had never seen before. He also noted the many peoples he met in South America. In December 1832, when the ship landed on Tierra del Fuego at the southern tip of the continent, Darwin was shocked by the local Fuegians. They wore few clothes and painted their faces, something the Englishman had never encountered.

Ideas that changed the world

Darwin wrote in his journal: "On the 19th of August we finally left the shores of Brazil. I thank God, I shall never again visit a slave-country. To this day, if I hear a distant scream, it recalls with painful vividness my feelings, when passing a house near Pernambuco, I heard the most pitiable moans, and could not but suspect that some poor slave was being tortured…"

Exploring the ideas

Darwin saw many different ways of life during the voyage. He was most shocked by the appalling conditions of the slaves he saw in Brazil in 1832. They had been brought from Africa to work on sugar plantations. Darwin thought slavery was barbaric and brought shame to anyone who allowed it to happen. Captain FitzRoy disagreed, and the two men had a furious argument about slavery. Slavery only became illegal in Britain the following year; in the United States, it was not banned until 1863.

After seeing slavery for himself, Darwin felt horrified and ashamed.

HISTORY'S STORY

The *Beagle*'s voyage was part of Britain's ambition to build a larger **empire**. One of the *Beagle*'s main tasks was to map coastlines to help strengthen the British Empire, which relied on sea transportation. The expedition was also intended to spread European civilization to peoples they encountered. Darwin's scientific work was only a small part of the mission.

MAKING OBSERVATIONS

Darwin had decided to sail on HMS *Beagle* to see as much of the world as he could and to learn from what he saw. While at sea, he read the first volume of Charles Lyell's *Principles of Geology*, which had been published in 1830. The book made a big impression on Darwin. Lyell argued that processes of nature, such as **erosion**, happen slowly but constantly, so must have been happening for millions of years to shape Earth.

In 1835, while he was in Chile, Darwin saw a volcano erupt and experienced an earthquake. He felt the ground shake and saw cracks open up in it. The earthquake got Darwin thinking. He had found bits of seashells on top of hills. How did they get there? It seemed that parts of the landscape must once have been underwater, but had later risen up. If the landscape was continually changing, living things must be constantly changing too, in order to survive in their environment. In Patagonia, southern Chile, Darwin identified two different rheas living in different parts of the country. How had these ostrich-like flightless birds changed into two different species?

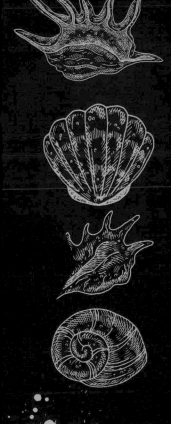

The lesser rhea was renamed "Darwin's rhea" for the naturalist.

Fossils are preserved in mud or sand, which turns to rock over millions of years.

Fossils everywhere

Another thing that puzzled Darwin was the many fossils he discovered in Argentina. The fossilized plants and animals he unearthed were similar to but not the same as plants and animals that were still alive. Why had they changed? One fossil was made up of bones as big as an elephant's, but it did not look like any animal he knew of that lived in South America. Darwin was convinced the bones came from an animal that had become extinct, or died out, but he did not know what kind of animal it was. He noted, "I have been wonderfully lucky with fossil bones. Some of the animals must have been of great dimensions."

Gifts for home

Darwin was so amazed by his huge fossil find that he insisted the crew help him bring it onto the ship. Later, he packed the bones up and sent them back to John Henslow in England. It was more than a year before Darwin learned if the fossils made it back safely. They did! All the bones arrived and Henslow showed them to leading scientists, who were amazed. Even though he was thousands of miles away, people in England were talking about Darwin and his discoveries!

KEY CLUES

On September 15, 1835, HMS *Beagle* arrived at its first stop in the Pacific Ocean, the Galápagos Islands. Darwin's visit to the islands became very famous, but at the time, it was not considered an important part of the trip. It was only later that Darwin realized the significance of what he had seen on the islands.

The Galápagos Islands had volcanoes and **lava fields** that Darwin wanted to study. He had not really thought about what kinds of wildlife he would find. Few people lived on the Galápagos but they were often visited by whaling ships that stopped to fish and hunt. The giant Galápagos tortoises were particularly prized because they could stay alive for months onboard a ship before being killed and eaten. The crew of the *Beagle* captured 18 tortoises for food. Darwin mistakenly assumed the giant tortoises had been imported from the Indian Ocean, so he did not give them much attention. In the same way, he believed marine iguanas, which were large, swimming lizards, had come from South America. In fact, they had not.

Darwin called the marine iguanas of the Galápagos "disgusting, clumsy lizards."

A casual approach

Darwin started to collect specimens of plants, animals, rocks, and fossils, as he did everywhere he visited. What amazed him most was how tame all the animals seemed to be. He even climbed on the back of a giant tortoise, which carried on walking as though he was not sitting on it! Darwin spent many days collecting samples, but did so in a slightly random way. He collected small birds, today known to be finches, on most of the larger islands in the archipelago, but did not bother to identify which bird came from which island. He assumed the birds were not related to each other. Later, he would regret his casual approach.

A helpful hint

The only permanent residents on the Galápagos Islands were prisoners who lived on what is now called Floreana Island. The governor of the prison mentioned to Darwin that he could tell which island a giant tortoise came from just by looking at the patterns on its shell. The comment was the starting point for some of Darwin's most important ideas. He started to pay more attention to the birds he was collecting.

The giant tortoises found on the Galápagos Islands live only there.

HISTORY'S STORY

The Galápagos Islands lie 600 miles (965 km) from South America in the Pacific Ocean. The 18 large islands and many smaller islands are scattered on either side of the equator. "Galápagos" comes from an old Spanish word for tortoises, a name first given to the islands by Spanish adventurers in the 1500s.

TIMELINE OF THE VOYAGE OF HMS *BEAGLE*

1831, December 27

The ship sets sail from Plymouth, England.

1832, January 16

The expedition makes its first stop, in St. Jago (now named Santiago), in the Cape Verde Islands. Darwin observes the landscape, noting the island's many valleys.

Darwin wondered if erosion caused by running water had carved the valleys of St. Jago.

1832–1833

As the *Beagle* sails up and down the coast of South America, Darwin goes onto the land. He is amazed by the Amazon rain forest and discovers fossils of extinct animals, including the giant sloth.

1835, January 19

The volcano of Mount Osorno, in Chile, erupts.

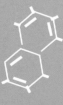

In the Andes Mountains, Darwin saw fossilized seashells thousands of feet above sea level, which further confirmed Lyell's theories.

1835, February 20

Darwin experiences an earthquake in Chile. He sees new land rising out of the sea, just as Charles Lyell had described.

1835, January–September

Darwin goes on an inland expedition across the Andes Mountains into Peru.

1835, September 15

The *Beagle* arrives in the Galápagos Islands for a five-week stay. Darwin collects finch specimens, which later prove his theory of evolution. Similarly, he notes that the design on the shells of giant tortoises varies from island to island.

1836, January

In Australia, Darwin starts to wonder about natural selection as he learns that the continent's **Indigenous** peoples are dying from disease while European settler Australians survive.

1836, October 2

HMS *Beagle* sails into Falmouth, England. Darwin brings back 1,529 species bottled in spirits or alcohol, 3,907 more labeled species, a 770-page diary, and journals of his trip.

CHAPTER 4

Darwin was fascinated by the similarities between humans and species such as chimpanzees.

FORMING A THEORY
Key Ideas

The basic ideas of the theory of evolution included:

a. Not all members of a species are exactly the same.

b. Small differences are passed from one **generation** to the next.

c. Only the fittest in a species survive.

d. Living things that are better adapted to their environment tend to survive and have more offspring. This process is known as natural selection.

e. Each generation **inherits** variations that are useful.

f. Over a long period, a species adapts so much it becomes a new species.

g. All species on Earth have adapted from common **ancestors** and are therefore related.

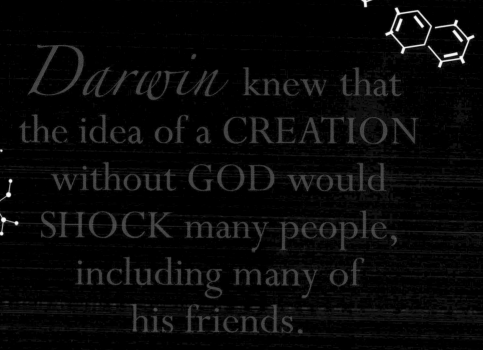

Darwin knew that the idea of a CREATION without GOD would SHOCK many people, including many of his friends.

WRITING UP THE VOYAGE

After four years, nine months, and five days, HMS *Beagle* returned to England on October 2, 1836. Having seen so many new things on his voyage, Darwin was eager to start work. He had collected thousands of specimens, which all needed to be classified and recorded before he could study them. There was too much for one man to do!

This watercolor portrait of Darwin was made in the late 1830s.

Once again, John Henslow came to the rescue. He arranged for experts to take specimens to study. A leading **ornithologist**, John Gould, examined the birds, while Richard Owen, who was a famous animal **anatomist**, examined the fossils. Darwin even got to meet his hero, Charles Lyell, whose writing had been so important on his travels. Much to Darwin's surprise, Lyell treated Darwin as a hero and the two soon became friends.

Although Darwin lived in Cambridge, he was spending much of his time in London visiting experts. So in March 1837, he moved to London, despite hating the dirty city. One day, John Gould came to see him. Gould told him the small birds he had brought from the Galápagos were wrongly labeled as blackbirds, gross-beaks, and finches. Gould pointed out that the birds were all finches, but that each one was a separate, unidentified species. This got Darwin thinking.

Ideas that changed the world

Darwin wrote in On the Origin of Species: *"Probably all the organic beings which have ever lived on this earth have **descended** from some one primordial form, into which life was first breathed by the Creator."*

Exploring the ideas

Darwin knew that his ideas challenged the foundations of the world he lived in. He was arguing that everything now living on Earth had shared a common ancestor, created by God, but had changed over time from what God had initially created. He was careful to suggest to his audience that God created the world. Darwin knew that going against the teachings of the Church would cause huge public disagreement, and he wanted to avoid that for as long as he possibly could.

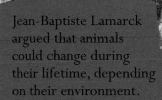

Jean-Baptiste Lamarck argued that animals could change during their lifetime, depending on their environment.

HISTORY'S STORY

Before Darwin came up with his theory of evolution, other scientists had formed theories of how life developed on Earth. Jean-Baptiste Lamarck came up with the idea of "transmutation," his term for evolution. Geologists like Charles Lyell proposed a theory called "uniformitarianism." They argued that Earth had been subject to ongoing, gradual changes over millions of years.

CHANGING SPECIES

Darwin later said that the two years after he returned to England on the *Beagle*, were the busiest time of his life. Not only did he have thousands and thousands of specimens to catalog, but he also had to write up the journal he kept on the voyage. In addition, he was not getting any younger. It was time to settle down, marry, and start a family.

Darwin's research was slowly but surely leading him to believe in what we now call a theory of evolution. (Darwin himself did not yet use the word "evolution"; he called it transmutation.) Charles Lyell had shown that Earth was much older than the Bible suggested. The many fossils Darwin had collected were proof that countless plants and animals that once lived were now extinct. However, some of the fossils had clear similarities to plants and animals that still lived.

Darwin agreed with Lamarck that plants and animals had changed over time. Unlike Lamarck, Darwin did not believe the changes took place within an individual lifetime and were then passed on to offspring. He thought the process must take much longer. As early as 1837, Darwin was at work on his own theories of how species changed. Taking the Galápagos finches as a starting point, he noted how each finch had a different beak. He wondered if one single species of finch had arrived on the islands from South America long ago. Over time, the birds' beaks had changed to be able to best eat the food available on their particular island.

Darwin studied the cactus finch, whose beak is suited to feeding on prickly plants.

Mother Nature

In October 1838, Darwin read *Essay on the Principle of Population* by Thomas Malthus. Malthus suggested that all animal mothers give birth to more offspring than can possibly survive. If all their offspring lived, the planet would soon be overrun. In fact, most animals died when they were very young. Reading the book was Darwin's "light bulb" moment. He saw that, in order to survive, an animal must have some advantage over other young animals. He had already concluded that transmutation of the species took place, but now he could see how. Over time, the tiny advantages that kept animals alive led to small changes that eventually grew large enough to create a new species.

Darwin's gray warbler finch eats insects.

The medium ground finch has a short, strong beak suited to eating hard seeds.

Charles and Emma (pictured as an elderly woman) were very happy, despite his initially scientific approach to the relationship.

HISTORY'S STORY

In 1838, Darwin married his cousin Emma Wedgwood. He took a scientific approach to the decision, writing out lists of the advantages and disadvantages. In 1842, they moved to Down House in Kent, where they lived for the rest of their lives. They had 10 children, three of whom died, including Darwin's favorite, Annie, who died at the age of 10.

FINDING PROOF

With a growing family, Darwin's London house was cramped. The family bought a country house south of London named for a nearby village. Life in Down House followed a regular routine. Darwin worked on his theories in the peace of his study, while Emma was kept busy with the children.

In 1843, Darwin finally finished his immense five-volume work, *The Zoology of the Voyage of H.M.S. Beagle*. It included experts' opinions on the different animals he described. Just a year later, a book called *Vestiges of the Natural History of Creation* was published by an **anonymous** author about the transmutation of species. Riddled with errors, it claimed humans were descended from animals, but did not give any explanation of how evolution happened. Many people assumed Darwin was the author, which made him furious! In fact, it was written by a man named Robert Chambers. The book's unpopularity convinced Darwin that he could not publish his theory of evolution until he had proven his ideas by studying a species in detail.

This drawing from *The Zoology of the Voyage of H.M.S. Beagle* shows a Bennett's chinchilla rat.

In 1844, Darwin wrote an essay containing his main ideas on evolution and sent it to his friend Joseph Dalton Hooker to read. Concerned about his own health, Darwin told Emma to publish the essay if he died. He knew his theory conflicted with the Bible's account of the Creation, and that the idea of a Creation without God would shock many people.

Proof of natural selection

Darwin needed proof to support his theory of natural selection. He needed to know how useful **adaptations** are passed down to the next generation. He spent eight years, between 1846 and 1854, studying barnacles. By the end of that period, he thoroughly hated barnacles, but the long and boring study had proved to be invaluable. He could now show how natural selection worked in practice.

Barnacles are invertebrates that attach themselves to rocks.

In 1855, the naturalist Alfred Russel Wallace wrote to Darwin to describe a theory about the origin of new species. Wallace's ideas were very similar to Darwin's, and Darwin's friends urged him to quickly publish his own findings. If Wallace published first, then Darwin's ideas would seem to be copying Wallace's and 20 years of work would be wasted. In fact, the men decided to publish their ideas together. On June 30, 1858, Darwin and Wallace's joint paper on the theory of evolution was presented to the Linnean Society. The society was the leading club for natural philosophers in London.

Alfred Russel Wallace believed Darwin had gathered more evidence than he had, and that Darwin's work would help support his own findings.

HISTORY'S STORY

Alfred Russel Wallace (1823–1913) was a Welsh naturalist who traveled to South America, Indonesia, and Malaysia, where he came up with his theory of evolution. He was never as famous as Darwin, but did not mind his friend getting the credit for much of their theory.

DARWIN'S KEY WORKS

The Zoology of the Voyage of H.M.S. Beagle (1838–1843)

This five-volume account of Darwin's trip was an instant hit on publication. It made him famous.

This drawing from *The Zoology of the Voyage of H.M.S. Beagle* shows the gray-bellied shrike-tyrant of South America.

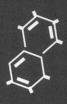

On the Origin of Species by Means of Natural Selection, or the Preservation of Favoured Races in the Struggle for Life (1859)

This book introduced the public to Darwin's theory of evolution, and forever altered beliefs about how life on Earth was created.

Fertilisation of Orchids (1862)

This was Darwin's first detailed demonstration of the power of natural selection to explain complex relationships. It looked at the way insects helped different orchids to **reproduce**.

The Descent of Man, and Selection in Relation to Sex (1871)

Darwin showed how humans were descended from earlier animals, and how sexual selection works separately from and alongside natural selection.

The Expression of the Emotions in Man and Animals (1872)

This was Darwin's fourth work on evolutionary theory, and among the first books ever to include printed photographs.

The Power of Movement in Plants (1880)

In this book, which covered how plant shoots grow toward light, Darwin gave more evidence for the theory of natural selection.

The Formation of Vegetable Mould through the Action of Worms (1881)

Darwin's last scientific book, published shortly before his death, explored the behavior of earthworms and their effect on soil.

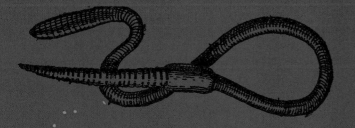

These photographs from Darwin's 1872 book show people expressing emotions.

CHAPTER 5

This statue of Charles Darwin stands in his hometown of Shrewsbury in England.

ON THE ORIGIN OF SPECIES
Key Facts

Full title: *On the Origin of Species by Means of Natural Selection, or the Preservation of Favoured Races in the Struggle for Life*

Publisher: John Murray

Published: November 24, 1859

First print run: 1,250 copies (sold out)

Number of editions in Darwin's lifetime: 6

U.S. edition published: January 1860

Number of languages translated into during Darwin's lifetime: 11

Money earned by Darwin: For 1st edition, £130; by 6th edition, £3,000

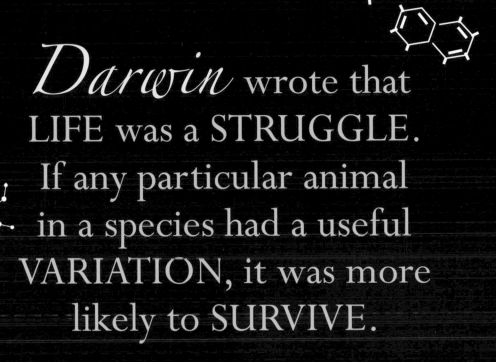

Darwin wrote that LIFE was a STRUGGLE. If any particular animal in a species had a useful VARIATION, it was more likely to SURVIVE.

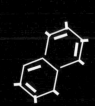

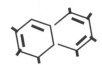

PUBLISHING HIS IDEAS

More than 20 years after Charles Darwin first started work on his theory of evolution, *On the Origin of Species* appeared in 1859. Aimed at the general reader rather than Darwin's fellow scientists, the book was an immediate hit. It sold out of its first print run, of just 1,250 copies. The book was so popular that a second print run of 3,000 copies was immediately prepared.

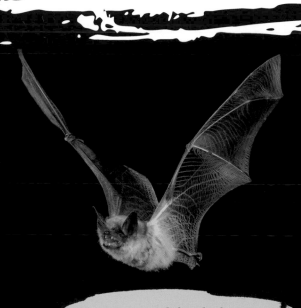

In 14 chapters, the book summed up Darwin's idea that all species share a common ancestor and that they are all related. He also introduced the reader to key ideas, such as natural selection, and the idea that life is a struggle that only the fittest survive. Darwin argued that a series of tiny changes over many generations can lead to the adaptation of a whole species and even the introduction of new species.

Darwin noticed that the bones in a bat's wing resembled those of the human hand.

... life is a struggle that only the fittest survive ...

Darwin was very careful to take his readers step by step, so that they did not misunderstand what he was telling them: that natural selection works over millions of years to bring about what he called the transmutation of species. To back up his arguments, he included lots of facts.

Ideas that changed the world

In On the Origin of Species, *Darwin wrote:"What can be more curious than that the hand of a man, formed for grasping, that of a mole for digging, the leg of the horse, the paddle of the porpoise, and the wing of the bat, should all be constructed on the same pattern, and should include the same bones, in the same relative positions?"*

Exploring the ideas

By listing animals that would appear to have no connection, Darwin suggested a groundbreaking idea to his readers. He explained that these animals are, in fact, all connected by having similar bone structures but for different purposes. He was demonstrating how species have evolved to adapt to their environment. However, for people living in Victorian England, this would have been shocking, because the Bible states that humans are unique in the world, quite different from animals.

Darwin believed it could not be coincidence that a mole's claws were similar to human hands.

HISTORY'S STORY

Darwin had been working on his theory of transmutation for more than 20 years, but ended up publishing *On the Origin of Species* very quickly after starting to write it. After he and Wallace presented their findings in 1858, Darwin set out to write his theories. He hoped the finished work would be around 30 pages—in the end, it was 502 pages!

45

CAUSING A STIR

Darwin was less concerned with *On the Origin of Species'* impressive sales figures than with just what people thought of his ideas. He knew that his ideas were groundbreaking and that many people, especially some of his fellow scientists and leading **theologians**, would disagree with them. Darwin was naturally quiet and shy. The idea that he might have to defend his ideas in public filled him with dread.

A close friend of Darwin's, the science professor Thomas Huxley, told Darwin he would do whatever it took to defend *On the Origin of Species* because Huxley believed it was so important. He wrote, "I am prepared to go to the stake...," meaning he would risk death. Other friends, such as Charles Lyell, praised the book. Darwin's brother Erasmus wrote to him: "I really think it is the most interesting book I ever read."

Thomas Huxley was the first person to suggest that birds evolved from dinosaurs.

Even Captain FitzRoy ... hated the book.

Some friends, however, were not impressed. Adam Sedgwick, who had first taught Darwin geology, was a deeply believing Christian. Sedgwick thought much of the book was false and that Darwin's ideas about evolution would send him to eternal **damnation**. Even Captain FitzRoy, who had invited Darwin onto the *Beagle*, hated the book so much he said he regretted inviting Darwin aboard.

"The most dangerous man in England"

Most of England's scientists and thinkers backed Darwin. They saw *On the Origin of Species* as a breakthrough in science and the first step in an exciting new journey to explain the **origins** of humankind and the planet. But many **devout** Christians were outraged. Darwin's ideas denied that God had created everything in the world as it still existed, so clearly suggested that the Bible's account of creation was wrong. One newspaper review called Darwin "the most dangerous man in England." The country was split between those who supported Darwin and his radical theories and those who thought his work was **blasphemous**.

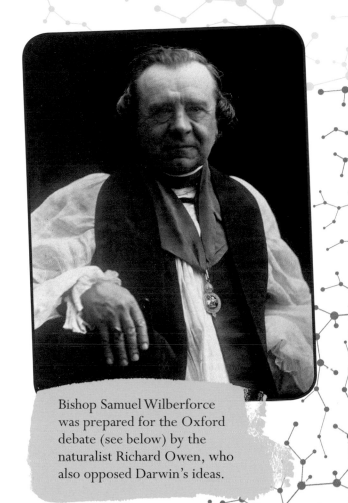

Bishop Samuel Wilberforce was prepared for the Oxford debate (see below) by the naturalist Richard Owen, who also opposed Darwin's ideas.

HISTORY'S STORY

In 1860, a famous debate took place in Oxford. In front of a large audience, Bishop Samuel Wilberforce defended God's role in creation, while Darwin's theories were defended by his friends Thomas Huxley and Joseph Hooker. The general opinion was that Huxley and Hooker won the argument. Their views were more in keeping with the public mood: by 1860, the idea of evolution was not new, and many people no longer believed the Bible was literally true.

END OF DAYS

Ever since his voyage on HMS *Beagle*, Charles Darwin had suffered from periods of ill health. The last few decades of his life were marked by sickness, but he continued to work on a number of books that furthered his theories. After the publication of *On the Origin of Species*, Darwin and his ideas became more popular.

More and more people came to accept Darwin's ideas, and his theories were spread yet farther by Thomas Huxley. Unlike Darwin, Huxley loved engaging with the public. He lectured to hundreds of working people who did not have the chance to read Darwin's ideas about the origins of humans. Many working people liked the idea that their low status in life was not permanently fixed by God.

… their low status in life was not permanently fixed by God.

Across Europe, Darwin's popularity grew as his works were translated into French, German, Russian, and many other languages. Despite his fame, Darwin avoided public events. He spent his time in Down House working on his books. His daily routine continued much as it had for decades.

This photograph shows Darwin as an old man, when he was suffering from poor health.

Selecting a mate

In 1871, 12 years after *On the Origin of Species*, Darwin published his most daring book, *The Descent of Man*. Darwin described how humankind had evolved from primitive ancestors. This idea was no longer shocking to everyone, because many people had begun to understand the link between humans and animals. The book also suggested that sexual selection, or the way animals choose a mate, is a driving force of evolutionary change. This theory explained natural features such as male peacocks' beautiful tails. These features helped males attract a mate. The males were then able to pass these desirable features down to their offspring, while males with smaller tails could not.

The peacock's spectacular tail feathers are a way for the male to attract female mates.

The end comes

In November 1877, Cambridge University gave Darwin an honorary doctorate. Darwin did not normally bother with awards, but this one was special: it proved that his theories had been accepted. In April 1882, Darwin passed away in Emma's arms. He was buried in Westminster Abbey, London, an honor reserved for the most important Britons.

HISTORY'S STORY

Much to his amazement, Darwin's last book, *The Formation of Vegetable Mould through the Action of Worms*, was the most popular of all his works, selling thousands of copies within the first few weeks. He had been convinced that nobody would buy it!

SPECIES AND PLACES NAMED AFTER DARWIN

Species
More than 300 species carry Darwin's name, including:

Darwin's rhea (*Rhea pennata*), found in Argentina, Bolivia, Chile, and Peru

Darwin's frog (*Rhinoderma darwinii*), found in Argentina and Chile

Darwin's finches (about 15 species), which helped with his theory of evolution, found in the Galápagos Islands, Ecuador

Ida (*Darwinius masillae*), a 47-million-year-old fossil **primate**, found in Germany

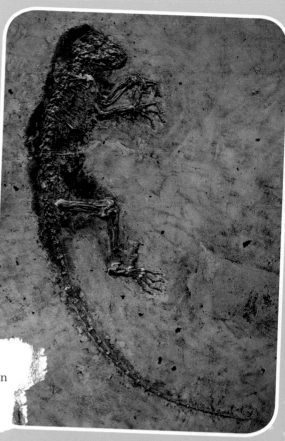

Ida was a lemur-like creature discovered in the year of Darwin's 200th birthday.

Geographical features
Around 20 places and institutions have been named after Darwin, including:

Charles Darwin University, Northern Territory, Australia

Cordillera Darwin, Tierra del Fuego, Chile

Darwin, East Falkland, Falkland Islands

Darwin, Northern Territory, Australia

Darwin College, Cambridge University, England

Darwin Crater, Mars

Darwin Crater, the Moon

Darwin Glacier, California, United States

Darwin Island, Danger Islands, Antarctica

Darwin Island, Galápagos Islands, Ecuador

Darwin Sound, British Columbia, Canada

Darwin Sound, Tierra del Fuego, Chile

Mount Darwin, Andes Mountains, Chile

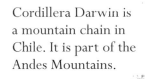

Cordillera Darwin is a mountain chain in Chile. It is part of the Andes Mountains.

CHAPTER 6

A CHANGED WORLD
Legacy

Scientists today understand the process of how parents pass on characteristics to their offspring.

Before Darwin: Most people believed God had created the world and everything in it as it currently existed.

After Darwin: Most people accepted that humankind developed over millions of years of evolution.

Before Darwin: Most people believed that all species were created at the same time by God.

After Darwin: Most people believed that species can become extinct or new species can be created as a result of evolution.

Before Darwin: Most people did not understand why some species were more widespread than others.

After Darwin: Most people accepted that natural selection explained why only those species best suited to their environment survive.

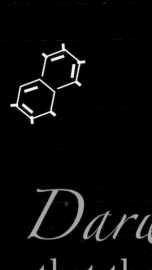

Darwin believed that the qualities that made him a successful SCIENTIST were a LOVE OF SCIENCE, great PATIENCE in considering his ideas, HARD WORK in gathering facts, and equal parts IMAGINATION and COMMON SENSE.

A GUIDING HAND

Today, Charles Darwin is hailed as the father of modern evolutionary **biology**. He gave scientists a method for examining all living things: We can study features of plants and animals, and figure out how they help its survival. Darwin's work means that scientists are able to trace evolution by looking at how fossils have changed over millions of years.

Darwin's work opened the way for many breakthroughs in learning about life, but there were also other important contributions. One came from a monk named Gregor Mendel (1822–1884), who lived in a monastery in what is now the Czech Republic. Around the time that Darwin was getting ready to publish *On the Origin of Species*, Mendel was growing pea plants to study how parents passed on features from one generation to the next.

Darwin's work opened the way for many breakthroughs in learning about life.

Darwin had not answered the key question of how features were passed on from parent to offspring. Mendel studied pea plants between 1856 and 1863. He discovered that each parent passes on exactly half the information required to shape its offspring. This takes place through chemical instructions we now call **genes**. Genes determine if an organism is big or small, what its colors will be, and many other characteristics.

Mendel studied the different colors of pea flowers to understand how characteristics are passed on.

54

Ideas that changed the world

In 1838, Darwin wrote in his notebook: "Man in his arrogance thinks himself a great work, worthy of the interposition of a deity. More humble, and I believe truer, to consider him created from animals."

Exploring the ideas

When Darwin first proposed his ideas on how humans had come into being, he knew that he was going against the Bible and that many people would think his ideas were blasphemous. At first, he tried to suggest that it was possible that God had created a world that had since evolved from its origins, but the more he studied the more uncertain he became. He declared himself to be an agnostic, meaning that he did not believe anything could be known of the existence of God.

This 1871 cartoon mocked Darwin by showing his head on a monkey's body.

HISTORY'S STORY

Today, many people are able to combine their religious beliefs with an acceptance that life has changed over time through evolution. However, some Christians insist that the biblical account of how God created the world is literally true. These people are often called Creationists, and their belief is called Creationism.

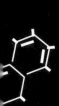

SOCIAL DARWINISM

Darwin's theory of evolution was about how plant and animal life has physically changed over time. Later, people applied his ideas to human society to explain why some people do better than others. The idea that Darwin's theories explained society became known as **Social Darwinism**, even though Darwin himself had nothing to do with it.

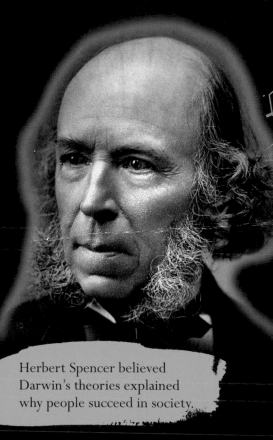

Herbert Spencer believed Darwin's theories explained why people succeed in society.

The phrase "survival of the fittest" was first used by scientist Herbert Spencer (1820–1903), after reading *On the Origin of Species*. Spencer used the phrase to explain why some people were more successful than others. He compared human society to the natural world and argued that, just as animals evolved in nature, so human society also evolved. Spencer was also one of the first people to use the word "evolution" in place of the word "transmutation."

Spencer believed that some people were destined to become rich, while others were destined to fail and have sad, poor lives. The "fittest" people were those who were smarter, worked harder, and had better health. Over time, the weaker members of society would die out, as their children would die young from hunger and disease. Only the "fittest" would survive. Spencer named his theory "Social Darwinism."

Some Social Darwinists used Darwin's theories to excuse cruel practices, such as child labor.

A cruel theory

Darwin was not happy to have his name attached to an idea that was not his and with which he did not agree, but the name and the idea caught on. Since Darwin's time, people have used it to justify the false idea that some races are naturally superior to other races. Some politicians have also used Social Darwinism to suggest that it is acceptable for the rich to make use of the poor. The American oil magnate John D. Rockefeller, one of the richest people in modern times, used Social Darwinism to justify growing rich on the efforts of workers: "The growth of large business is merely the survival of the fittest. It is merely the working out of a law of nature."

Today, many people believe that Social Darwinism has wrongly been used to justify cruelty by suggesting that society is naturally unfair. They argue that humans have a duty to care for the less fortunate in society. Just because somebody is poor, they say, does not mean they deserve to be poor.

UNDERSTANDING THE WORLD

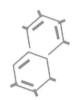

What are the next steps in evolution? Today, scientists understand far more about the biological processes that create evolution. Advances in our understanding of the chemicals that make up life, together with the ongoing discovery of fossils from the distant past, are helping biologists learn more about what life was once like on Earth. They also help them to predict how it might develop in the future.

When Darwin was researching his theories, few fossils of ancient humans and their ancestors had been found. Darwin guessed that humans had descended from apes, but modern scientists have proof that he lacked. In the past 60 years, fossil discoveries have proven that modern humans, the species *Homo sapiens*, evolved in East Africa between 200,000 and 1 million years ago. Even older fossils have proven a link between early humans and apes.

Almost a century after Darwin published *On the Origin of Species*, scientists discovered the way parents pass on features to their offspring through a **molecule** called deoxyribonucleic acid (DNA), which is in almost every **cell** of every living thing. In 1953, the scientists James Watson and Francis Crick discovered that DNA was very long, and resembled a ladder twisting in a spiral. DNA carries all the instructions for how living things form and grow.

Analysis shows that humans and chimpanzees share 96 percent of their DNA!

Still evolving

Early in the 2000s, there are still many unanswered questions about evolution. One of the most important is whether humans are still evolving—and if so, into what?

Medicine has wiped out many deadly diseases, but humans are still plagued by illness. People live longer, which has led to an increase in cases of cancer and age-related illnesses. Humans have not yet evolved to avoid these killers. In the same way, as Earth's climate continues to change, humans have not evolved sufficiently to deal with extreme heat or cold. We are continually shaping and changing our environment, but what impact will that have on our bodies? Some scientists believe that humans might be able to use robotics and **artificial intelligence (AI)** to evolve in completely new ways.

This reconstruction shows us what Lucy (see right) looked like. She took us one step closer to understanding where we came from.

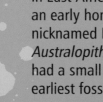

HISTORY'S STORY

In 1974, archeologists made an exciting discovery in East Africa: the 3.2-million-year-old fossil of an early hominid, or human-like creature, they nicknamed Lucy. Lucy belonged to a species called *Australopithecus afarensis*, walked on two legs, but had a small ape-like skull. At the time, Lucy was the earliest fossil found of an upright-walking hominid.

BEYOND DARWIN

Darwin's work has inspired biologists right up to the present day. These are some scientists who have made important contributions to our developing understanding of Darwin's theory of evolution:

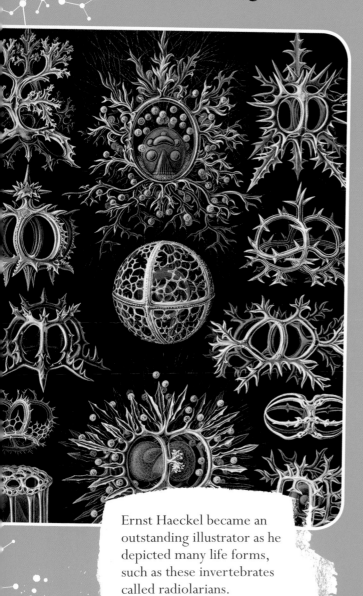

Ernst Haeckel became an outstanding illustrator as he depicted many life forms, such as these invertebrates called radiolarians.

Gregor Mendel (1822–1884)

The monk from Bohemia proved how traits are passed from parents to offspring.

Ernst Haeckel (1834–1919)

The German scientist looked at how babies form inside their mothers, while also studying other, simpler forms of life besides plants and animals. He argued that humans had evolved from simple life forms.

William Bateson (1861–1926)

The first person to use the term **genetics**, Bateson translated Mendel's work from German to English. He disagreed with Darwin's theory that species adapt through slow, gradual changes, instead arguing that changes could come about in sudden jumps.

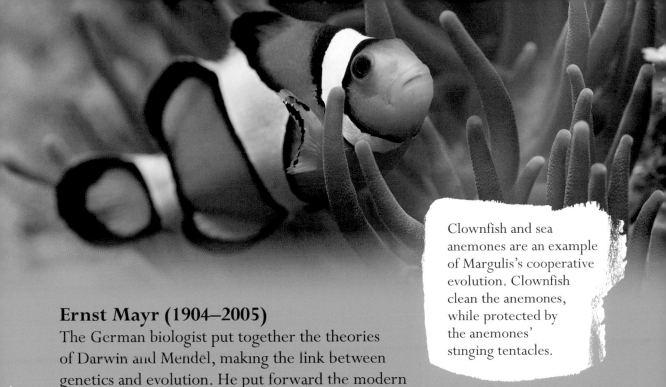

Clownfish and sea anemones are an example of Margulis's cooperative evolution. Clownfish clean the anemones, while protected by the anemones' stinging tentacles.

Ernst Mayr (1904–2005)

The German biologist put together the theories of Darwin and Mendel, making the link between genetics and evolution. He put forward the modern definition of a species: a species is a group of living things that can produce offspring only with each other.

Lynn Margulis (1938–2011)

Margulis revolutionized the way evolutionary scientists understood how natural selection works. She showed that species could evolve because of cooperation with each other as well as competition.

Stephen Jay Gould (1941–2002)

The American biologist suggested that evolution was made up of very long periods when nothing happened, followed by shorter periods of frantic activity when many changes took place.

Richard Dawkins (1941–)

A British biologist, Dawkins has put genes at the heart of evolutionary study, and has written a lot about belief in God being opposed to reason.

GLOSSARY

adaptations The processes by which a species changes its features to better fit its environment

anatomist Someone who studies the structure of the body

ancestors Relatives who lived in the past

anesthetic A substance that causes lack of feeling or awareness to dull pain for surgery

anonymous Not named

archipelago A group of islands

artificial intelligence (AI) Programming that allows machines to "think" like humans

astronomer Someone who studies space

bacteria Microscopic one-celled living things that live in soil, water, air, and on bodies. Some can cause disease.

barbaric Cruel or primitive

biology The study of living things

blasphemous Showing disrespect to God

botany The study of plants

cell A tiny building block for all living things

classification Sorting things into related groups

damnation Eternal punishment suffered in hell

data Facts

descended Related to a particular ancestor

devout Having deep religious feeling

divine God-like

economist Someone who studies how money, trade, and industry are organized

empire A group of countries controlled by a single ruler

erosion The wearing away of rocks and soil by wind, water, and other natural processes

evolution The slow change of species over generations as they adapt to their environment

fossils The remains of ancient animals or plants that have been preserved in stone

generation The members of a species born around the same time

genes Chemical instructions that pass from parents to their offspring

genetics The study of how characteristics are passed on from one generation to the next

gentlemen Men of good social position, wealth, and often noble birth

geologist Someone who studies rocks

geology The study of rocks

Indigenous The original inhabitants of an area, or the First Peoples

Industrial Revolution The period between 1760 and 1840 when new inventions led to the development of machine tools and factories

inherits Receives something from one's parents

invertebrates Animals without internal skeletons

lava fields Areas covered in hardened lava

mate A sexual partner

molecule A small chemical unit made up of atoms

natural historian Someone who studies all aspects of the natural world

natural philosophy The studies that we now call natural history

natural selection Part of the theory of evolution stating that living things that adapt well to their environment are more likely to survive and pass on their characteristics to their young

naturalist A person who studies animals and plants

offspring The young of any living organism

organism Any living thing

origin The point or place where something begins

ornithologist Someone who studies birds

pioneering Leading the way

primate A member of a group that includes humans, monkeys, apes, and their close relatives

Prussian Someone from the old German state of Prussia, which came to dominate German politics

reproduce To create young

sexual selection The way animals choose their mate, based on their preference for particular characteristics

Social Darwinism An argument that suggests that "survival of the fittest" can be applied to society

species A group of animals that has evolved to be different from any other animals and can produce offspring only with each other

specimens Things collected by a scientist for study

theologians People who study religion

theory A scientific principle suggested to explain facts and observations

FOR MORE INFORMATION

BOOKS

Anderson, Margaret J. *Charles Darwin: Genius of a Revolutionary Theory* (Genius Scientists and Their Genius Ideas). Berkeley Heights, NJ: Enslow Publishers, Inc., 2015.

Byrne, Eugene and Simon Gurr. *Darwin: A Graphic Biography.* Washington, DC: Smithsonian Books, 2013.

Krull, Kathleen. *Charles Darwin* (Giants of Science). New York, NY: Puffin Books, 2015.

Sullivan, Laura L. *Charles Darwin: Groundbreaking Naturalist and Evolutionary Theorist* (Great Minds of Science). Minneapolis, MN: Core Library, 2015.

WEBSITES

Biography—www.britannica.com/biography/Charles-Darwin
A website with information about Darwin's life.

Evolution—www.pbs.org/wgbh/evolution
Part of PBS's *Evolution* series, this website offers information about all aspects of evolution and Darwin's life.

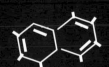

Galápagos Islands—www.darwinfoundation.org
Information about the Galápagos Islands and the Charles Darwin Research Station based there.

Timeline—http://darwin-online.org.uk/timeline.html
A detailed timeline of Darwin's life.

INDEX